The Ten Square Dance

Early Childhood Mathematics

Karen Hegarty

Balboa Press books may be ordered through booksellers or by contacting:

Balboa Press
A Division of Hay House
1663 Liberty Drive
Bloomington, IN 47403
www.balboapress.com
1 (877) 407-4847

ISBN: 978-1-4525-2243-2 (sc)
ISBN: 978-1-4525-2244-9 (e)

Library of Congress Control Number: 2014916786

Printed in the United States of America.

Balboa Press rev. date: 10/13/2014

BALBOA PRESS
A DIVISION OF HAY HOUSE

Dear Parents/Teachers,

The aim of this book is to help children understand the link between an amount of objects and the symbol (numeral) we use to represent that amount.

As you read out the question at the beginning of each page, it will not take long for your child/children to start answering. If the answer is incorrect, you can say "Good try", and then give them the correct answer. As you say the numeral you can trace it with your finger and then to help them make the connection you can count the chickens on the opposite page starting from the bottom.

Counting from the bottom of the grid helps children to understand that numbers grow. I have illustrated the chickens within a ten square grid so that subconsciously your child is beginning to put the numbers within a decimal context.

Rhyme, repetition and colour stimulate the memory centres of the brain and of course it is just plain FUN!

By pairing up chickens side-by-side children also have a frame of visual reference in which to understand odd and even numbers. I have included two grids in the back of the book and included the phrases that I have used to help this understanding. Don't rush this process though, as it takes a wee while to make those first links.

To deepen the learning, and to engage kinaesthetic learners (those who learn by doing) I have also included two pages of circular illustrations that can be cut out, laminated and have numerals **1** through to **10** written on the back. Add magnetic tape, draw a grid on a magnetic board and you have a wonderful game you can use in conjunction with the book, that children love to play over and over.

Want to go a step further? Make a grid out of a cut off end of your yoga mat and use toy cars; dinosaurs; figurines or pretty much whatever your child is interested in. Take maths outside and draw a grid in the sand/dirt and use any natural resources. Most of all have fun and HAPPY MATHEMATICS LEARNING my friends.

With love
from Karen

Karen Hegarty BTchLn (Early Childhood Education)

How many chickens at the ten square dance?

One little chicken, an odd man out, all he can do is jump and shout!

How many chickens at the ten square dance?

2

Two little chickens, wing to wing.
Two little chickens, chirp and sing.

How many chickens at the ten square dance?

3

Three little chickens, on the floor,
waiting for more chickens
coming through the door

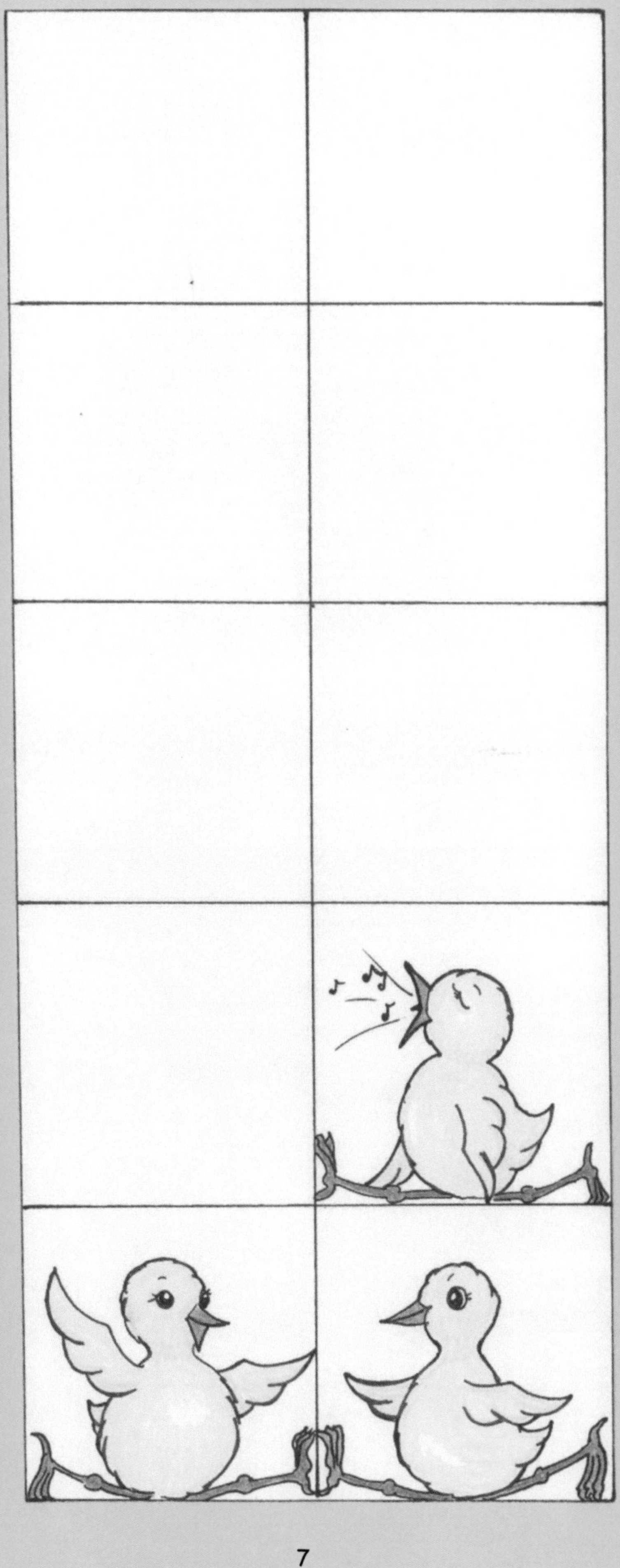

How many chickens at the ten square dance?

4

Four little chickens, toe to toe.
Dance to the music, nice and slow.

How many chickens at the ten square dance?

5

Five little chickens boogie down,
serious dancing, done with a frown.

How many chickens at the ten square dance?

6

Six little chickens, I declare!
Six little chickens with bows in their hair.

How many chickens at the ten square dance?

Seven little chickens, not so big.
Little yellow chickens do an Irish jig.

How many chickens at the ten square dance?

8

Eight little chickens with mighty hoots,
dance all night in red gumboots.

How many chickens at the ten square dance?

9

Nine little chickens, move and shake.
How much noise can a little chicken make?

How many chickens at the ten square dance?

10

Ten little chickens, take a bow.
All we can see are tail feathers now.

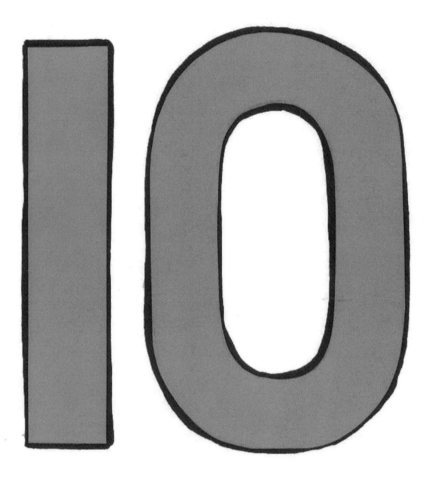

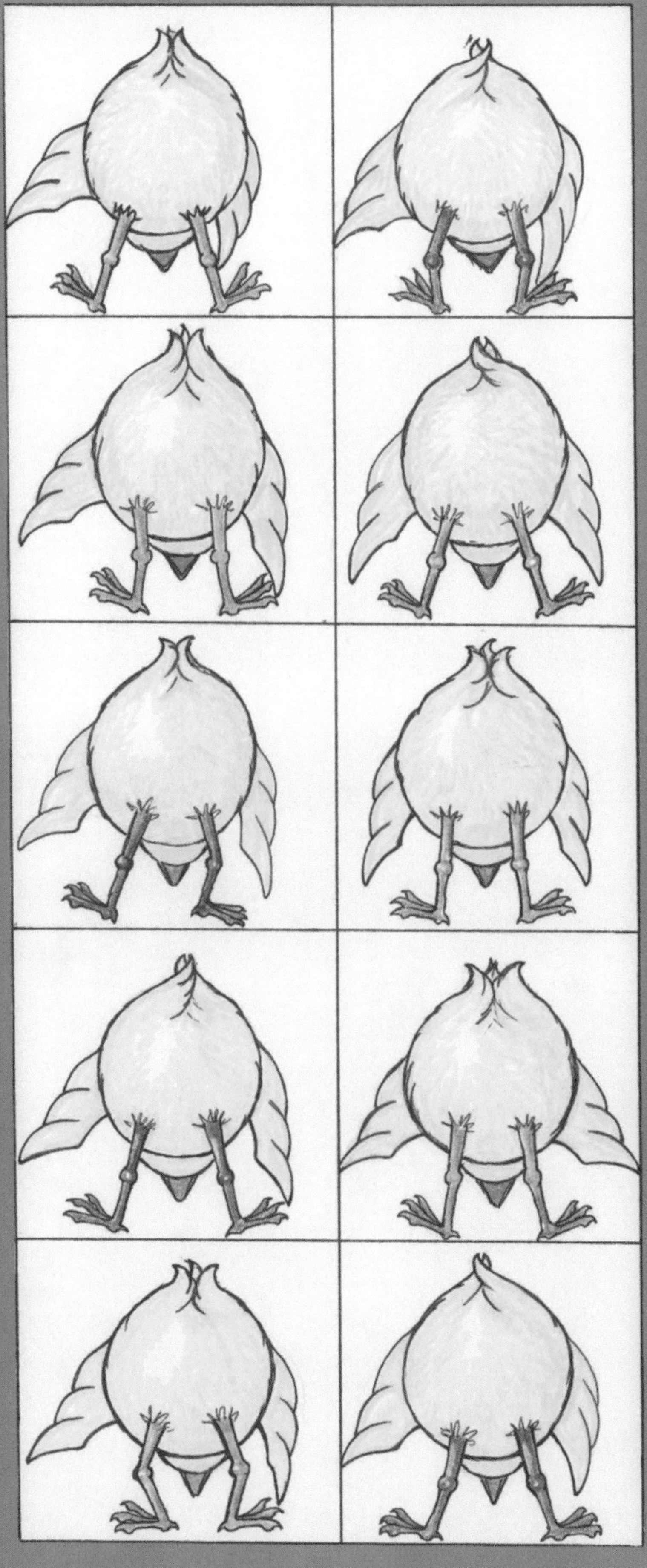

1 2 3 4 5

6 7 8 9 10

11 12 13 14 15

16 17 18 19 20

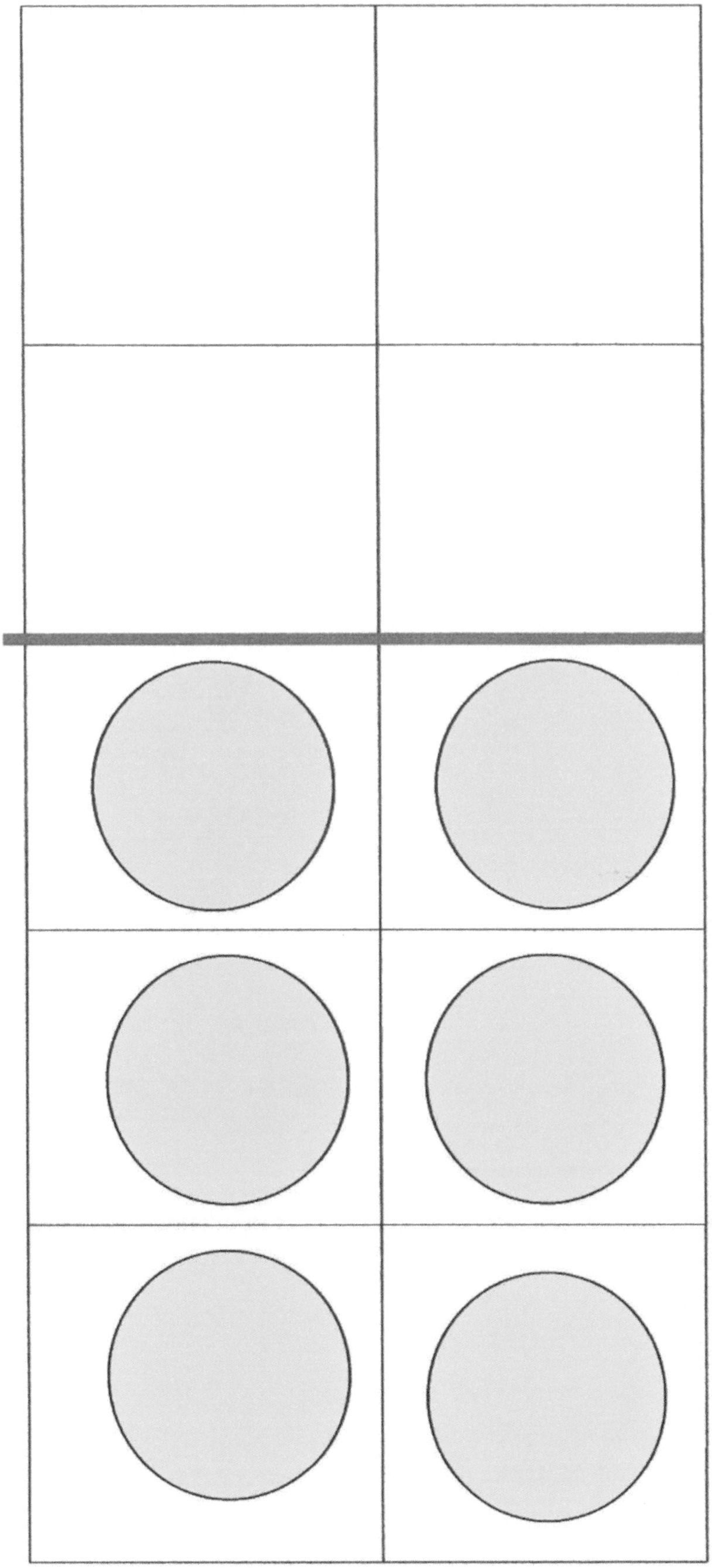

Even numbers are flat and even across the top.

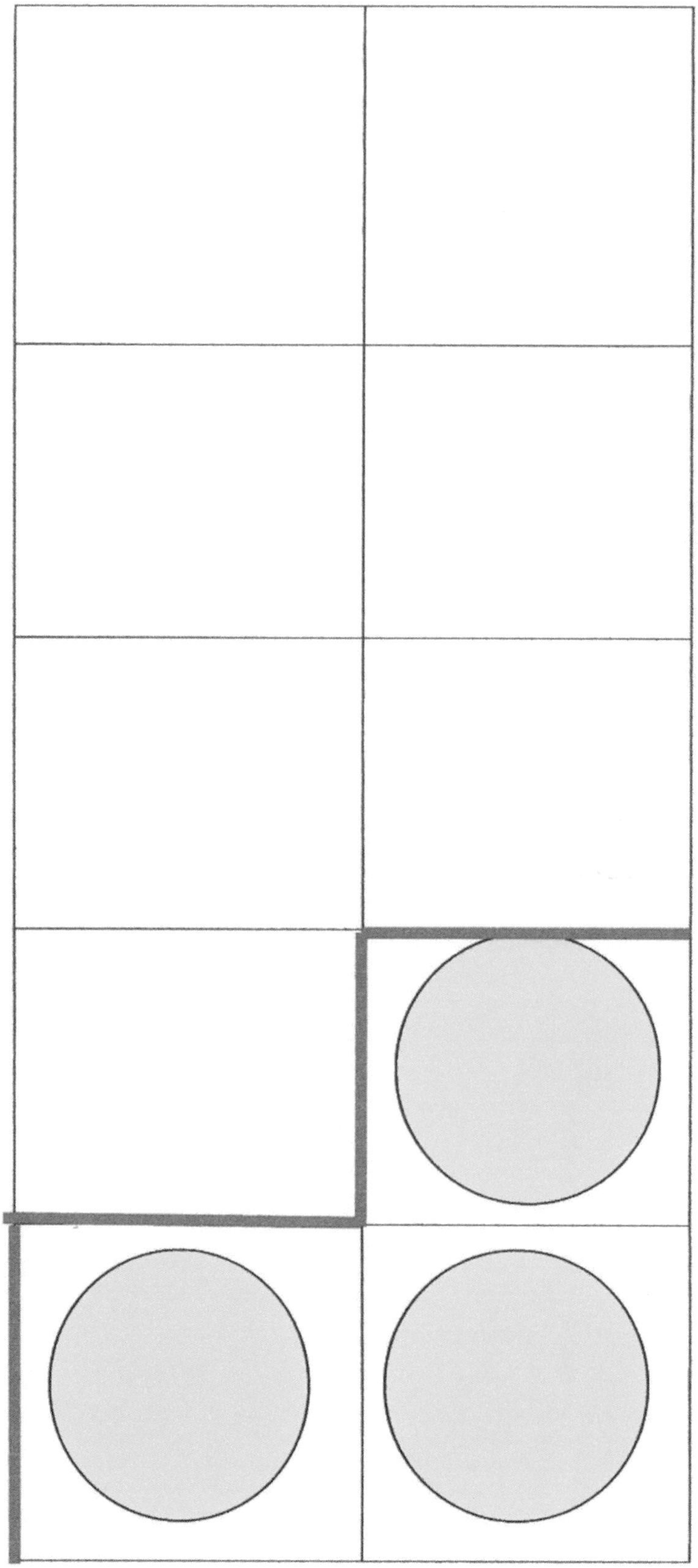

Odd numbers are like climbing up stairs.

Lightning Source UK Ltd.
Milton Keynes UK
UKIC02n0849041114
1953UKAU00004B/5